FEDERAL EMERGENCY MANAGEMENT AGENCY

UNITED STATES FIRE ADMINISTRATION

NATIONAL FIRE DATA CENTER

An NFIRS Analysis:

Investigating City Characteristics

and Residential Fire Rates

Prepared by:

TriData Corporation
1000 Wilson Boulevard
Arlington, Virginia 22209

This publication was produced under Contract EMC-95-C-4717 by TriData Corporation for the United States Fire Administration, Federal Emergency Management Agency. Any information, findings, conclusions, or recommendations expressed in this publication do not necessarily reflect the views of the Federal Emergency Management Agency or the United States Fire Administration.

APRIL 1998

TABLE OF CONTENTS

Executive Summary

The objective of this study was to identify relationships between city characteristics and residential fire rates. The study analyzed data from 27 cities reporting to the United States Fire Administration's National Fire Incident Reporting System (NFIRS). NFIRS is the largest fire data set in the country, and each year almost one million new records are added. For each city, fire rates for eight different categories of fire cause were studied, as well as the overall level of fires. The causes included fires due to arson, children playing, careless smoking, cooking, heating, electrical distribution, appliances, and open flames. In seeking to explain city-to-city variation in fire rates, we examined climate, age structure of the population, and differences in the socioeconomic status of city residents. The findings of this study are presented in comparison with the findings of previous analyses.

Among the major findings of this study are:

- Particular city characteristics were found to be strongly related to fire rates. The most common factors related to higher fire rates were climate and the age of the housing stock. Cities with worse climates and older housing stocks had a greater likelihood of fire.

- Five of the eight causes of fire were found to be strongly related to at least one city characteristic. These included fires due to arson, children playing, careless smoking, heating, and electrical distribution. Much of the variation between cities in the rates of these fires could be explained by factors not controllable by the fire service.

- Cooking fires were not found to be strongly related to city characteristics. This was unexpected because other studies have found strong links between poverty and the incidence of cooking fires. The use of cities as the unit of analysis may explain why no significant correlates of cooking fires were identified in this study.

The intent of research such as this is to help identify and clarify relationships between characteristics of people and places and fire risk. This information can be used for a variety of purposes, including the design, targeting, and evaluation of fire prevention programs. For example, cities with high proportions of children under age five need to recognize that their risk of children playing fires is higher than in other cities. They should compare their progress in reducing the rate of these fires against cities with similar proportions of children. Similarly comparisons could be made for other causes of fires.

Introduction

This report identifies relationships between city characteristics and the causes of residential fires, with special emphasis on climate, demographic, and socioeconomic factors.[1] It is the second in a series studying factors that have been linked to increased risks of fire in the United States. The first report, entitled *Socioeconomic Factors and the Incidence of Fire,* surveyed the fire literature from the past twenty years on the link between socioeconomic factors and the incidence of fire in U.S. cities.[2]

Why are the causes of home fires of special interest? First, fires in residential structures account for the vast majority of civilian fire deaths and injuries each year. Second, many of these fires are eminently avoidable. A high proportion of all fires that occur in residential structures are directly attributable to human activities. For the years 1993-1995, over half (55 percent) of home fires were caused by cooking, arson, open flames, careless smoking, and children playing with fire.[3,4] If heating fires are included, the percentage of all home fires attributable to human activities or carelessness rises to 73 percent, or almost three-fourths of fires.[5]

Studying the association between city fire rates and factors such as climate, socioeconomic characteristics, and demographic characteristics should lead to a clearer understanding of circumstances that are conducive to fire. Similarly, research can inform the way the fire service targets and evaluates prevention efforts. Since human activities are directly implicated in a high proportion of all home fires, policy and education interventions stand out as the most effective means to significantly reduce the number of fires in many communities. Fewer fires in turn mean fewer deaths, fewer injuries, and less property loss.

---

[1] "Residential" as the term is used here refers to structure fires only.

[2] United States Fire Administration, Federal Emergency Management Agency, 1996. This report can be downloaded from the Fire Administration's web site at http://www.usfa.gov/nfdc/fius.htm.

[3] Arson is a legal term rather than a category used in NFIRS. In NFIRS, the specific circumstances of fires are analyzed, and the fires are grouped into different cause categories. One of these groups is fires of "incendiary or suspicious" origins. For purposes of brevity, these fires are referred to as arson fires in this report.

[4] These percentages are based on fires of known cause.

[5] Heating fires often result from human carelessness. Too often, fires originating in chimneys, fireplaces, or woodstoves are due to lack of maintenance or misuse of equipment, such as keeping combustibles too close to heat sources.

Review of the Literature

There is little cities can do about their climates or the demographic and socioeconomic characteristics of their residents. The fire service has even less influence on these factors. But prior research has shown that these characteristics can be useful in predicting the magnitude and nature of fire problems in different neighborhoods and sometimes in different cities. This type of information can be useful to cities and fire departments interested in targeting fire service and fire prevention education resources. Over the course of the last twenty years, researchers have used a variety of different approaches to measure the relationship between city characteristics and fire problems.

Most of the seminal studies linking fire to demographic and socioeconomic characteristics were conducted and published in the late 1970s and early 1980s. Since that time, only a limited amount of new research has been published. However, two recent studies, one by Jennings (1996) and one by the Statistics Unit of the New South Wales [Australia] Fire Brigades (1997), confirm many of the findings of earlier studies, most importantly that demographic and socioeconomic characteristics can be powerful predictors of community fire problems.

The importance of social and economic conditions for understanding community fire risk has been shown using a multitude of approaches. Table 1 displays the research strategies for several major studies, including the specific type of data analyzed, the fire indicators studied, and the indicators found to be significantly related to fire.[6]

---

[6] For a more detailed discussion of these studies, see the previously cited report *Socioeconomic Factors and the Incidence of Fire,* United States Fire Administration, Federal Emergency Management Agency, 1996.

Table 1. Summary of Major Demographic and Socioeconomic Studies on Fire Incidence

Study	Unit of Analysis	Fire Indicator	Significant Demographic and Socioeconomic Factors*	Comments
Munson, 1976	the five boroughs of New York City	fire rate for all structures	population density (+) median household income (-) percent of households with incomes under $5,000 (+)	While all the relationships exhibited linearity, statistical significance could not be reached because there were only five data points.
Schaenman, Hall, Jr., Schainblatt, Swartz, and Karter, 1977	census tracts in four cities, one county	fire rates for all buildings	parental presence (-) poverty (+) under-education (+)	While the complete list of variables and the amount of variation explained by each differed by city, poverty consistently ranked as one of the most powerful predictors of fire rates.
Karter, Jr., and Donner, 1978	census tracts in five cities	fire rates in residential (one- and two-family) dwellings	*Population characteristics:* family stability (-), poverty (+); *Housing characteristics:* crowdedness (+), home ownership (-), vacancy rates (+)	The variables with the most explanatory power varied by city, but in three out of five cities, poverty ranked first.
Gunther, 1981	census tracts divided into five groups by income and race	fire rates for a) all residential fires and b) each cause of fire	family income (-)	Income was a particularly powerful predictor of overall fire rates and rates of arson, careless smoking, cooking, and children playing fires.
Munson and Oates, 1983	a) 54 large U.S. cities, b) 36 NJ cities, and c) census tracts in Charlotte, NC	fire rates in buildings; in Charlotte, the dependent variable was residential fires.	income (-) poverty (+) home ownership (-) unemployment rates (+) African American population (+)	There was a generally a high degree of consistency across the data sets with different units of analysis. High levels of interdependence (multicollinearity) were evident among the independent variables, but the authors found that their results remained materially the same after a control for income was added.

* A plus sign (+) indicates that as the value of the demographic or socioeconomic factor increases, so does the fire rate. A minus sign (-) indicates that there is an inverse relationship between the demographic or socioeconomic factor and the fire rate: as one increases, the other decreases and vice versa.

Study	Unit of Analysis	Fire Indicator	Significant Demographic and Socioeconomic Factors*	Comments
Fahy and Norton, 1989	50 U.S. cities	fire rates	poverty (+)	Cities were classified by percent of population below poverty. Median fire rates for groups of cities rose as their proportions of people living in poverty increased.
Hall, Jr., 1993	communities	fire rates	community size (curvilinear)	Small and large communities have the highest fire rates. Medium-sized communities have the lowest fire rates. This is related to concentrations of poverty in small and large communities.
Jennings, 1996	census tracts	fire rates for residential structure fires	median household income (-) percent of population less than 17 or older than 64 (+) percent of female-headed households with children (+) percent of vacant units (+)	Only these four variables were included in the final regression model to avoid problems of high correlations among the independent variables.
Statistics Unit, New South Wales Fire Brigades	post code areas	number of total fires, house fires, structure fires, arson fires, and bush and grass fires	age (-) educational attainment (-) income (-) unemployment (+) home ownership (-)	Socioeconomic and income measures were particularly strong predictors of house and structure fire rates.

* A plus sign (+) indicates that as the value of the demographic or socioeconomic factor increases, so does the fire rate. A minus sign (-) indicates that there is an inverse relationship between the demographic or socioeconomic factor and the fire rate: as one increases, the other decreases and vice versa.

The studies by Jennings and the New South Wales Fire Brigades were published in the 1990s. The Jennings study was described in the earlier U.S. Fire Administration report, so it is not reviewed in depth here. The study by the New South Wales Fire Brigades, however, is more recent and was not reviewed previously.[7] The New South Wales study examined fire rates in post code areas (like U.S. zip codes) in Sydney, Australia. Researchers used correlation analysis, factor analysis, and cluster analysis to study five different fire indicators: the number of total fires, house fires, structure fires, arson fires, and bush and grass fires. They identified six factors of post code areas that were significantly related to the incidence of fire. These were age, ethnicity, educational attainment, income, unemployment, and home ownership.

Unlike most previous research, the New South Wales study included one fire indicator representing the specific cause of arson. Of the other studies listed in Table 1, only Gunther (1981) researched relationships between specific causes of fires and socioeconomic characteristics. Analyzing census tracts within Toledo, Ohio, for USFA Gunther found that income was a strong predictor of rates for certain categories of fire cause, particularly arson, careless smoking, cooking, and children playing fires.

Using factor analysis, the New South Wales research found that two factors explained a moderate amount of variation in arson fire rates. The authors characterized the first factor as reflecting "demographic and social climate" conditions within post codes. These conditions included education levels, rates of home ownership, population size, and age structure of the population. The second factor reflected "socioeconomic/income structure" conditions within post codes. This factor included strong effects from income, unemployment levels, ethnicity of residents, and the prevalence of low-skill jobs.[8]

Methodology

The objective of the current study is to contribute to the understanding of the relationship between city characteristics and residential fire rates by analyzing 1993-1995 data from the National Fire Incident Reporting System (NFIRS). NFIRS is the largest fire data set in the United States.

---

[7] Statistics Unit, Corporate Strategy Division, New South Wales Fire Brigade. *Socio-Economic Characteristics of Communities and Fires.* NSW Fire Brigades Statistical Research Paper, June 1997.

[8] New South Wales Fire Brigade, 1997, p. 5.

Each year almost one million new fires are added to NFIRS. The U.S. Fire Administration, with the assistance of the National Fire Information Council, maintains the data system. Annually, fire departments from all over the country report on the number and types of fires to which they have responded. The system is a voluntary one, but it is estimated that over half of the nation's fire departments participate in NFIRS and close to half of the fires attended by fire departments are reported to it – providing a very large, robust sample.

The largest U.S. cities and counties participating in NFIRS are the best candidates for undertaking a city-level statistical analysis. The reason is that larger fire departments have historically been more likely to have sophisticated data management systems. These systems, in turn, report very high proportions of all their fire incidents to NFIRS. These high levels of reporting allow researchers to be confident that they have an accurate assessment of the number and types of fires that occur within participating cities or counties.

This research analyzed the total number of residential structure fires reported by 27 major U.S. cities and counties and the number of those fires in each locality attributable to arson, children playing, careless smoking, heating, cooking, electrical distribution, appliances, and open flames.[9] These represent the eight leading causes of fire in the U.S., and together they account for over 90 percent of all residential structure fires. Table 2 lists the 24 cities and three counties included in the NFIRS data set. For each city, the data set included the rate of all fires and rates for each of the eight fire causes listed above.[10] City fire rates were calculated by dividing the aggregate number of residential fires in each category by city population.

---

[9] Originally, 30 metropolitan areas were included in the data set. After examining the initial data set, two cities, Kansas City, Kansas and Detroit, Michigan and one county, Orange County, California were eliminated due to data considerations.

[10] Because most of the places included in the data set were cities, the term "city" is used in several places throughout this report as shorthand for "city or county".

Table 2. U.S. Cities and Counties in the Study Data Set

Baltimore City, MD	Montgomery County, MD
Baltimore County, MD	Nashville, TN
Boston, MA	New Orleans, LA
Buffalo, NY	Norfolk, VA
Cleveland, OH	Oklahoma City, OK
Columbus, OH	Portland, OR
Dallas, TX	Rochester, NY
Denver, CO	San Antonio, TX
El Paso, TX	San Diego, CA
Fort Worth, TX	San Francisco, CA
Houston, TX	Virginia Beach, VA
Jacksonville, FL	Washington, DC
Los Angeles County, CA	Worcester, MA
Memphis, TN	

A separate data set was created containing a variety of climate, demographic, and socioeconomic indicators for each city and county. These indicators included total population, annual precipitation, race, income, poverty, family structure, age of housing, etc. These data items were extracted from the *1994 City and County Data Book* CD-ROM. This data set was then matched with the NFIRS data set created above. The result was a single data set with fire, climate, demographic, and socioeconomic data for each city and county.

The final data set was imported into SPSS version 6.1.3 for analysis. Predictions about the relationship between city characteristics and city fire rates were tested by computing correlation coefficients and performing multiple regression analysis. Definitions of the indicators that were significantly related to particular fire causes are discussed in the text below. A complete list of the city characteristics included in the analysis and their definitions is included in Appendix A.

Analysis

The objective of this analysis was to determine whether city characteristics were useful in predicting a) overall residential fire rates for localities and b) rates of residential fires attributable to specific causes. One limitation of studying fire rates among a diverse set of U.S. cities is the difficulty in controlling for local conditions. Studies that concentrate on single areas implicitly control for such factors as climate, population trends, and the history of the local housing stock. In the current study, the difficulty of controlling for these factors introduced the possibility of findings that would be more complex to explain.

Since the analysis used aggregated fire and city characteristics, the findings are interpreted as identifying factors associated with increased residential fire rates at the city level. For example, the results showed that in cities with relatively low levels of median income, arson fire rates tended to be higher. It was not possible using this data set, however, to link the occurrence of fires to the demographic or socioeconomic characteristics of particular households. This would have required an analysis using households or perhaps neighborhoods as the unit of analysis rather than cities.

It was expected that city demographic and socioeconomic variables would be linked to the frequency of cooking, arson, children playing, and careless smoking fires because these types of fires are directly linked to human activities. Specifically, it was expected that the rates of these fires would be higher in poorer communities and communities with significant indicators of decline, such as high unemployment rates. These communities tend to be characterized by lower median household incomes, greater proportions of people living in poverty, and higher proportions of female-headed households.

Studies such as those reviewed earlier in this report have often found a link between poverty and poverty-related indicators and increased fire rates. Further evidence of this relationship through this study will reinforce the need for fire prevention efforts in certain cities, particularly efforts focused on teaching citizens how to properly handle fire and fire-related materials and about their responsibility for practicing fire-safe behaviors.

In contrast to fires directly related to human activities, it was expected that the rate of fires in cities due to heating, electrical distribution, appliances, and open flames would occur more randomly and show less sensitivity to city demographic or socioeconomic factors. For example, to the extent that heating, electrical distribution, and appliance fires are caused by mechanical malfunction, then the socioeconomic characteristics of cities should not be especially helpful in predicting fire rates for those causes.

While open flame fires are related to human activities, it was expected that there would not be a strong relationship between open flame fire rates and city characteristics. Open flame fires are most often associated with candles, matches, and lighters in residential structures. In this data set, it was not expected that demographic or socioeconomic factors would strongly predict open flame residential fire rates among cities.

The expected relationships between city fire rates and city characteristics are listed in Table 3. These predictions are based largely on Gunther (1981), the other study that has analyzed several individual causes of fires.

Table 3. City Characteristics and Fire Rates in Residential Structures

Residential Structural Fire Rate	Expected Strength of Relationship to City Characteristics
Overall Fire Rate	High
Arson Fire Rate	High
Children Playing Fire Rate	High
Careless Smoking Fire Rate	High
Cooking Fire Rate	High
Heating Fire Rate	Moderate
Electrical Distribution Fire Rate	Low
Appliances Fire Rate	Low
Open Flame Fire Rate	Low

Findings

Correlation Analysis

Correlation matrices revealed that 18 city characteristics were strongly related to at least one of the nine types of fire rates investigated.[11] These factors included the age distribution of the population, unemployment rates, median income, poverty levels, housing unit characteristics, housing tenure, housing costs, education, and household structure.[12] The correlation coefficients appear in Appendices B and C. Appendix B shows how each of the socioeconomic, demographic, and climate factors included in the

---

[11] The correlation coefficients for these variables had significance levels of $P \leq .05$. $P \leq .05$ indicates that there is no more than a five percent chance that the two variables are correlated by chance only and are not in reality related to one another.

[12] Two interaction terms, "race x poverty" and "race x education", were investigated. Neither term was significant in any of the regression equations.

analysis related to one another. Appendix C shows how causes of fires were related to each of the socioeconomic, demographic, and climate factors.

Multiple Regression Analysis

Multiple regression analysis was used to identify the city factors that explained the greatest amount of difference in residential fire rates between cities.[13,14,15] The results of the regression analyses are presented in Table 4. The table lists each city fire rate investigated followed by the city characteristics used in the final regression model. The order the factors entered the model is also presented. A regression "model" is the term used to represent all the factors included in a specific regression analysis and resulting equation. For each factor in the model, the direction of its association with the dependent variable is indicated, as well as its level of significance. Climate, demographic, and socioeconomic factors were indeed related to city fire rates. Where positive relationships were found, fire rates increased as the value of specific city characteristics increased. Conversely, negative relationships suggested that fire rates decreased as the value of specific city characteristics increased.

Column 3 lists the proportion of the difference in city residential fire rates explained by the regression model. The asterisks in column 3 indicate the significance level of the model.[16]

Two important socioeconomic factors were not found to be strongly related to residential fire rates because they were highly interrelated to other factors included in the regression models.[17] These factors were the percent of female-headed households and education, the latter being defined as the percent of persons over age 25 with a high school education. Education level, for example, was highly interrelated with population change, unemployment rate, and median income. The latter three factors were among the

---

[13] Stepwise regressions were run in SPSS version 6.1.3.

[14] The stepwise regression feature of SPSS was used to identify those independent variables that explained the greatest amount of variation in each dependent variable while controlling for all other independent variables in the regression equation.

[15] A factor analysis including all the independent variables significantly related to the dependent variables was conducted. The results did not improve on the multiple regression results with the original independent variables. This may be a function of the level of aggregation of the data used here. For factor analysis applied to fire rate research, see Statistics Unit, New South Wales (1997).

[16] The significance levels of the regression equations were evaluated using F-tests.

[17] The stepwise regression process retained those factors that explained the highest proportions of differences in fire rates between cities.

socioeconomic factors found to be useful in explaining differences in city fire rates (see
Table 4).

Table 4. Correlates of Fire Causes with City Characteristics

(1)	(2)	(3)
Residential Structural Fire Rate	**Significant City Characteristics (in Descending Order)**[18,19]	**Percent of Difference in Fire Rate Explained by the Model**
Overall Fire Rate	1) Annual precipitation (+) ** 2) Percent pre-1940 housing units (+) ** 3) Percent of population under age 5 (+) **	64% ***
Arson Fire Rate[20]	1) Median Income (-) ** 2) Percent rental housing (+) *	70% **
Children Playing Fire Rate[21]	1) Percent change in population, 1980-1992 (-) *** 2) Percent of population under age 5 (+) ***	60% ***
Careless Smoking Fire Rate	1) Percent pre-1940 housing units (+) **	40% **
Cooking Fire Rate	None	-
Heating Fire Rate	1) Annual precipitation (+) *** 2) Percent rental housing (-) *	49% **
Electrical Distribution Fire Rate[22]	1) Annual precipitation (+) ***	62% ***
Appliance Fire Rate	None	-
Open Flame Fire Rate	1) Percent pre-1940 housing units (+) *	27% *

Column 2 reports T-test significance levels; Column 3 reports F-test significance levels. The symbols are as
follows:
*** indicates P<.001   ** indicates P<.01   * indicates P<.05

Discussion

The results of this analysis showed that certain city characteristics were
significantly related to residential fire rates for seven out of the nine causes of fires. The

---

[18] Using stepwise regression analysis.

[19] Unless otherwise indicated, the regression results were estimated using Weighted Least Squares (WLS)
regression. An examination of Ordinary Least Squares (OLS) regression residuals (the difference between
actual and predicted values) plotted against predicted values of each dependent variable revealed apparent
heteroscedasticity. WLS was used in an attempt to improve each model's fit with the data.

[20] This variable was logged to adjust for a positive skew in its distribution.

[21] OLS regression was used to estimate this model. WLS did not improve upon the fit of the OLS regression
equation.

[22] Initially, race was significant in the regression model. However, because of the close correlation between
race and income levels, the model was re-run controlling for income. In this model, the only significant
variable was annual precipitation.

amount of the difference explained by city characteristics ranged from 70 percent for arson fires to 27 percent for open flame fires. City characteristics explained at least half of the variation in the case of overall city fire rates and for four of the seven different causes of fires: arson, children playing, careless smoking, and electrical distribution.

Using these results, the expectations presented at the outset of this study can be evaluated. Table 5 includes the original expectations about the usefulness of city characteristics for predicting residential fire rates (from Table 3). The last column presents the study's findings. The expectation that city characteristics would be "highly" predictive of fire rates was confirmed in three cases, for overall city fire rates, arson fire rates, and children playing fire rates.

Table 5. Expected versus Actual Results for City Characteristics and Fire Rates in Residential Structures

Residential Structural Fire Rate	Expected Predictive Power (Old Studies)	New Research Findings on Predictive Power	Proportion of Difference in Fire Rate Explained by City Characteristics
Overall Fire Rate	High	High	64%
Arson Fire Rate	High	High	70%
Children Playing Fire Rate	High	High	60%
Careless Smoking Fire Rate	High	Moderate	40%
Cooking Fire Rate	High	Low	-
Heating Fire Rate	Moderate	Moderate	49%
Electrical Distr. Fire Rate	Low	High	62%
Appliance Fire Rate	Low	Low	-
Open Flame Fire Rate	Low	Low	27%

Similarly, city characteristics were confirmed to be "moderate" predictors of heating fire rates and "low" predictors of appliance and open flame fire rates.

Surprisingly, after controlling for median income, the model for electrical distribution fire rates explained a "high" amount of the difference in fire rates between cities. This finding was somewhat unexpected since the connection between human activities and electrical distribution fires is generally considered to be less direct than for arson or children playing fires, for example. However, the electrical distribution category includes overloaded sockets, worn-out lamp cords, and other behavior-related fire hazards, not just faulty wiring in buildings.

Contrary to expectations, city characteristics did not prove to be powerful predictors of cooking fire rates. In the stepwise regression analysis, none of the climate, demographic, or socioeconomic factors entered the model.[23] This finding differs significantly from Gunther's findings in the city of Toledo. One possible explanation is that differences in rates of cooking fires due to socioeconomic or other factors that are detectable at the census tract or neighborhood level may be masked by other factors at the city level.

Similarly, while "high" predictive power was expected for careless smoking fire rates, the results here suggest that the strength of the relationship with city characteristics is better characterized as "moderate".

To put the findings of this research into perspective, the results are similar to those of other researchers. The 1977 Urban Institute study was able to explain 60 percent of fire rate variation among census tracts within cities. In two recent studies, Jennings (1996) was able to explain up to 83 percent of the variation in fire incidence rates among Memphis, Tennessee census tracts using socioeconomic indicators. The New South Wales Fire Brigades (1997) analysis was able to explain up to 82 percent of variation in fire incidence among greater Sydney post codes.

Results as strong as those of Jennings and the New South Wales Fire Brigades are more likely when studying smaller geographical areas. By focusing on smaller areas, variables such as climate and the histories of populations and building stocks that are difficult to control for statistically among geographically diverse places are naturally controlled for in single area studies.

One of the socioeconomic factors included in this study was "poverty", defined as the proportion of the population in each city living below the federal poverty line. Interestingly, and contrary to what was expected from the literature, this variable was not significant in any of the regression analyses conducted for this study. However, this was due to the composition of the data set. Specifically, there were several southern cities in

---

[23] This case meant that none of the variables improved the probability of the F-statistic of the regression model by .05 or more.

the data set with high poverty levels, but relatively low fire rates. Similarly, there were three northern cities that had relatively low poverty rates, but high fire rates.

The exception was for arson fire rates and children playing fire rates. The poverty measure was significantly related to both of these causes of fires. However, in the regression analyses, poverty was not retained as significant in the final regression models because other factors explained more of the variation in rates between cities. However, as is discussed below, many factors closely related to poverty, such as median income and the proportion of renter households, were significantly related to fire rates.

In the section below, each of the fire rates in this study is discussed in light of the climate, demographic, and socioeconomic variables that were found to be significantly related. All fire rates pertain to residential structure fires only.

Overall Residential Fire Rates

The overall residential fire rate reflects the rate of fires from all causes. As indicated in Table 4, three factors were significantly related to overall fire rates and together explained 64 percent of the variation between localities. These factors were ***annual precipitation, age of the housing stock*** (percent of housing stock built before 1940), and ***percent of population under age five***. These three variables are indicators of city climate, socioeconomic characteristics, and demographics, respectively. In other words, cities with bad weather, older housing, and lots of children under five can be expected to have more fires than sunny cities with older populations and relatively new housing. Old northern cities versus newer southern cities illustrate some of these differences.

The first important predictor of city fire rates was annual precipitation. This is one indicator of climate, and the cities in the data set with high amounts of annual precipitation tended to be northern cities where there are, on average, more heating days each year. (Appendix D lists each city in the data set and a percentage distribution of fires by cause). The more heating days, the greater the opportunity for a fire to occur as people use various heating devices to keep warm. Colder weather also means people spend more time indoors, and many of their activities, such as cooking, increase fire risk. Year after year, more fires occur in the winter months than in any other season.

While annual precipitation is a climate factor, in this data set it also identified a set of cities with similar characteristics. In this sense annual precipitation factor was an indicator of the age of cities, identifying whether they were older industrial cities or newer suburban-style cities with lower building and population densities and newer building stocks. In general, more of the older industrial cities are in decline, whereas many of the newer cities have growing economies. Residential fire rates were generally higher in the older industrial cities.

Related to the age of the city is the age of its housing stock. Age of housing stock was significantly related to overall residential fire rates even after accounting for the influence of annual precipitation. The overall residential fire rate tended to be higher in cities with older housing stocks and lower in cities with newer housing stocks. It is likely that newer housing is built to higher building codes, with better heating systems, and electrical systems better equipped to handle modern day appliances and electrical loads.

The third factor related to overall city residential fire rates was the percent of population under age five. It is unclear what the exact nature of the relationship between young children and increased overall fire rates is, especially given that high proportions of very young children were not correlated with any of the poverty indicators in the data set. However, as will be shown below, having more young children in households increases the risk of children playing fires. More children may also increase the risk of other types of fires by distracting adults. Of other recent fire studies, the New South Wales research found that the presence of young children was linked to higher fire rates. Also, Jennings found that increases in the percent of the population under 17 and over 64 were positively associated with higher fire rates.

Arson Fire Rates

At the outset of this study, it was expected that socioeconomic indicators would explain a high amount of the variation in residential arson fire rates among cities. The results of the multiple regression analysis strongly supported this hypothesis. Two socioeconomic factors, ***median income*** and ***proportion of rental housing units,*** explained 70 percent of the difference in arson rates among the localities in the data set, and the relationship was statistically very significant.[24] Median income was strongly negatively related to arson rates, accounting for more than four-fifths of the total explained variation.

---

[24] The regression equation had a significance level of P≤.0001.

In other words, as city median income fell, the rate of arson increased. Rental housing was positively related to arson rates, so as the proportion of rental housing increased, so did arson rates, even controlling for income. This reflects the fact that cities in the data set with more rental housing also tended to have older housing stocks and higher proportions of people living in poverty.

Children Playing Fire Rates

Two demographic variables, **percent change in population from 1980 to 1992** and **percent of population under age five,** explained 60 percent of the variation in residential children playing fire rates between cities. Population change is a demographic indicator of local economic health. People are attracted to cities with healthy economies where job opportunities are plentiful. Conversely, cities with declining economies often suffer population losses as people move to other locations. In this data set, population change was negatively related to children playing fire rates. Cities with higher levels of population growth had lower rates of children playing fires. The reverse was true for cities with low or negative population growth. Cities with declining populations also tended to have higher proportions of elderly residents, higher proportions of children, higher poverty rates, older housing stocks, more female-headed households, and fewer high school graduates.

Not surprisingly, the relationship between percent of population under age five and children playing fire rates was positive. Increases in the proportion of the population under age five meant higher rates of children playing fires. This was true even controlling for median income. In short, the more kids, the more children playing fires.

Careless Smoking Fire Rates

There was a moderate relationship between residential careless smoking fire rates and city characteristics, with one socioeconomic factor explaining 50 percent of the difference in fire rates. This factor was age of the housing stock. Age of housing stock was positively related to careless smoking fire rates. Other studies have indicated that lower income groups have higher proportions of smokers. The correlation coefficients calculated for this study revealed that the age of a locality's housing stock was significantly and positively related to higher poverty, higher unemployment levels, more female-headed households, lower education levels, and more rental housing. These relationships between socioeconomic status, age of housing, and the number of smokers may explain why communities with older, presumably lower quality, housing also had higher careless smoking fire rates.

Cooking Fire Rates

As indicated in Table 4, none of the city characteristics included in this study explained a significant amount of the difference in residential cooking fire rates between cities.[25] This was an unexpected finding. In his study of Toledo, Gunther (1981) found that cooking fires were significantly higher in low-income neighborhoods than in wealthier neighborhoods. According to *Fire in the United States*, cooking fires are the leading cause of residential structure fires in the nation. The lack of association between city characteristics and cooking may suggest that city-level fire data is too broad a measure to successfully detect variations in cooking fire rates associated with income, poverty, or other socioeconomic factors.

Heating Fire Rates

City characteristics explained a moderate amount of variation in heating fires (49 percent). This was expected since different climates require more heating use than do others. In addition, heating fires can be caused by equipment failure as well as by human activities. The two variables significantly related to residential heating fire rates were ***annual precipitation*** and ***percent of rental housing.*** Annual precipitation (which includes rain and snow) was positively related to heating fire rates. This is a function of the fact that localities with higher annual precipitation in the data set tended to be northern cities with more heating days and thus higher risks of heating fires. The proportion of rental housing was negatively related to heating fire rates, a finding that supports assertions that rental units, particularly those in apartment buildings, tend to have more satisfactory heating systems, often central heating that is professionally maintained.[26] With adequate central heating, residents are less likely to turn to alternative, less fire-safe heating devices such as wood stoves, kerosene heaters, and space heaters.

Electrical Distribution Fire Rates

City characteristics explained a high proportion of variation in residential electrical distribution fire rates. The first multiple regression run suggested that race was associated

---

[25] This finding did not come as a complete surprise. Initial research on this topic using proportional distributions of fires by cause indicated that cooking fires might relate differently than other categories to socioeconomic indicators.

[26] Gunther, 1981, p. 58.

with electrical distribution fire rates. However, because of the close association between race and socioeconomic status in the U.S., the analysis was re-run controlling for median income, even though income was not identified as a significant factor in the initial model. The results showed that ***annual precipitation*** was the only factor significantly related to differences in rates of electrical distribution fires. This model explained 62 percent of the variation between cities.[27]

Appliances Fire Rates

Similar to cooking fires, climate, demographic, and socioeconomic variables did not explain a significant amount of variation in residential appliance fire rates between cities. This finding was similar to Gunther's findings from Toledo. In his study, appliance fires showed little variation between income groups. Gunther explained that fires caused by equipment malfunctions were higher in the case of appliance fires than for any other equipment-related fire cause. Because equipment malfunctions should occur relatively randomly across households, indicators such as socioeconomic factors were not expected to be significantly correlated to appliance fire rates.[28]

Open Flame Fire Rates

City characteristics were not significant predictors of residential open flame fire rates. While the results showed that the rate of open flame fires increased as the age of the housing stock increased, this factor explained barely 25 percent of the variation in open flame fire rates between cities.

Conclusion

This research suggests that city characteristics are related to residential fire rates in important ways. The specific city characteristics of importance – whether climate, demographic, or socioeconomic – varied by the type of fire investigated. Significant relationships with these factors were identified for overall fire rates and the rates of arson, children playing, careless smoking, heating, and electrical distribution fires.

However, not all the expectations of this study from previous research were confirmed. In particular, this study did not find any significant relationships between city

---

[27] The second model explained more variation in electrical distribution fire rates and was more significant than the model that included race. The final model was significant at the P≤.0001 level.
[28] Gunther, 1981, pp. 57-58.

characteristics and cooking fire rates. This finding may be a result of conducting a city-level analysis, rather than studying particular fires or fire rates at the neighborhood level. Further research into the relationship between demographic and socioeconomic characteristics and cooking fires is needed to resolve this issue.

As in previous research, the relationships identified here between city characteristics and fire rates were not always straightforward. In the case of heating fires, the rate of fires decreased as the percent of rental housing increased, even though rental housing was correlated with poverty. This finding is similar to Gunther's (1981) finding that neighborhoods in Toledo's inner city had lower heating fire rates than other low income groups. Gunther and others attribute this to the presence of apartments, including public housing, with professionally installed and maintained central heating systems.[29]

This study confirms earlier research findings (Jennings [1996], New South Wales [1997]) that the percent of the population under five years of age is positively correlated to increased fire rates. This variable was significant in the models for overall fire rates and children playing fire rates.

Similarly the importance of building stock characteristics was evidenced by the significance of the age of housing stock variable. This variable was significant in the models for overall fire rates and careless smoking fire rates. The age of the housing stock is most likely an indicator of housing quality in neighborhoods where this housing is inhabited by relatively poor people. Where housing quality is lower, fire risk tends to increase. This increase can likely be traced to two related factors: the activities of

residents and the contents and structure of these housing units. Where mattresses are involved, for example, a dropped cigarette will ignite an old one more readily than a newer, post-1974 mattress that meets today's mattress flammability standards.

This analysis found no evidence that race contributes to fire rates independent of its effects through income. This conclusion concurs with the findings of Gunther (1981), Jennings (1996), and the New South Wales (1997) analysis. Likewise, as in the New South Wales study, the proportion of elderly residents was not found to be significantly associated with fire rates, regardless of fire cause.

---

[29] Gunther, 1981, p. 58.

The intent of research such as this is to help identify and clarify relationships between characteristics of people and places and fire risk. This information can be used for a variety of purposes, including the design and targeting of fire prevention programs. For example, the findings of this study suggest that cities with high proportions of children under age five should recognize that their risk of children playing fires is higher than in other cities. Each year in the U.S. over 25,000 house fires start as a result of children playing with matches or lighters – information as to what types of cities are at greater risk can be used to intervene and prevent these wholly avoidable fires.

REFERENCES

Gunther, Paul. 1981. "Fire-Cause Patterns for Different Socioeconomic Neighborhoods in Toledo, OH." *Fire Journal.* Vol. 75 (May), pp. 52-58.

Fahy, F. Rita and Alison L. Norton. 1989. "How Being Poor Effects Fire Risk...." *Fire Journal.* January/February, pp. 28-36.

Hall, John R. Jr. 1993. *The U.S. Fire Problem Overview Report through 1982: Leading Causes and Other Patterns and Trends.* Quincy, MA: National Fire Protection Association.

Jennings, Charles R. 1996. *Urban Residential Fires: An Empirical Analysis of Building Stock and Socioeconomic Characteristics for Memphis, Tennessee.* Unpublished doctoral dissertation.

Karter, Jr., Michael J. and Allan Donner. 1978. "The Effect of Demographics on Fire Rates." *Fire Journal.* Vol. 72, no. 1 (January), pp. 53-65.

Munson, Michael J. and Wallace E. Oates. 1983. "Community Characteristics and the Incidence of Fire: An Empirical Analysis." In *The Social and Economic Consequences of Residential Fires.* Chester Rapkin, Ed. Lexington, MA: D.C. Heath and Co.

New South Wales Fire Brigades. 1997. "Socio-Economic Characteristics of Communities and Fires.*" New South Wales Fire Brigades Statistical Research Paper,* ISBN 0 7310 3460 . Issue 4/97 (June 1997).

Schaenman, Philip et al. 1977. *Procedures for Improving the Measurement of Local Fire Protection Effectiveness.* Boston: National Fire Protection Association, pp. 53-71.

United States Fire Administration. 1993. *Fire in the United States, 1983 - 1990.* Washington, D.C.: Federal Emergency Management Agency, United States Fire Administration.

Appendix A. City Characteristics Included in the Analysis

The definitions and variable labels of each of the climate, demographic, and socioeconomic indicators included in this study appear below.

Independent Variables (27)	Variable Label
Population, 1992	Pop 1992
Population, 1980	Pop 1980
Population, percent change 1980-1992	%C80-92
African American population, 1990	B_pop90
Percent African American, 1990	%black
Percent one-person households, 1990	%one_hh
Median household income, 1989	med_hinc
Percent of persons below poverty level, 1989	%poverty
Percent renter-occupied housing units, 1990	%hu_rent
Labor force, percent change 1980-1990	%c_labor
Annual precipitation	rainfall
Average daily temperature in July	avg_temp
Percent of households receiving public assistance, 1989	%hh_asst
Percent of housing units built 1939 or earlier, 1990	%hu_1939
Unemployment rate, 1991	unemp_rt
Percent of female-headed family households	pct_fhfh
Percent of persons 25 and older who are high school graduates	pct_educ
Percent of population under age 5	under5
Percent of population aged 5 through 17	betw5_17
Percent of population 65 years and over, 1990	%elderly
Median value of specified owner-occupied housing units, 1990	med$_ooh
Median gross rent of specified renter-occupied housing units, 1990	med_rent
Percent condominium of occupied housing units, 1990	@condo
Female civilian labor force participation rate, 1990	@f_labor
Percent of year-round housing units vacant	vacancy
Percent of housing units with 5 or more units	@_5 units
Percent of year-round housing units with 1.01 or more persons per room	crowding
Dependent Variables (9)	**Variable Label**
Overall fire rate	fr_ntotl
Arson fire rate	fr_nars
Children playing fire rate	fr_nchld
Careless smoking fire rate	fr_nsmok
Heating fire rate	fr_nheat
Cooking fire rate	fr_ncook
Electrical distribution fire rate	fr_nelec
Appliance fire rate	fr_nappl
Open flame fire rate	fr_nopen

Appendix B.   Correlation Coefficients for the City Characteristics

The pages below contain correlation coefficients for the independent variables in the analysis, or those 18 city characteristics that were statistically significant with one or more of the fire indicators at the P≤ .05 level.

Independent Variables

-- Correlation Coefficients --

	POP_1992	POP_1980	@C80_92	@ELDERLY	B_POP90	@BLACK	@ONE_HH	MED_HINC	@HH_ASST	@POVERTY	@HU_1939
POP_1992	1.0000 -23 P= .	0.9760 -23 P= .000	-0.0391 -23 P= .859	-0.2008 -23 P= .358	0.6254 -23 P= .001	0.0302 -23 P= .891	-0.0346 -23 P= .875	0.0644 -23 P= .770	-0.1891 -20 P= .425	0.0102 -23 P= .963	-0.3886 -23 P= .067
POP_1980	0.9760 -23 P= .000	1.0000 -23 P= .	-0.2283 -23 P= .295	-0.0681 -23 P= .758	0.7455 -23 P= .000	0.1989 -23 P= .363	0.0992 -23 P= .652	-0.0386 -23 P= .861	-0.0460 -20 P= .847	0.1165 -23 P= .597	-0.2511 -23 P= .248
@C80_92	-0.0391 -23 P= .859	-0.2283 -23 P= .295	1.0000 -23 P= .	-0.6858 -23 P= .000	-0.4541 -23 P= .030	-0.6254 -23 P= .001	-0.6104 -23 P= .002	0.6120 -23 P= .002	-0.5974 -20 P= .005	-0.6231 -23 P= .001	-0.6032 -23 P= .002
@ELDERLY	-0.2008 -23 P= .358	-0.0681 -23 P= .758	-0.6858 -23 P= .000	1.0000 -23 P= .	0.0441 -23 P= .841	0.1678 -23 P= .444	0.5352 -23 P= .009	-0.3344 -23 P= .119	0.6408 -20 P= .002	0.2298 -23 P= .292	0.6927 -23 P= .000
B_POP90	0.6254 -23 P= .001	0.7455 -23 P= .000	-0.4541 -23 P= .030	0.0441 -23 P= .841	1.0000 -23 P= .	0.7538 -23 P= .000	0.3395 -23 P= .113	-0.1635 -23 P= .456	0.1993 -20 P= .400	0.2867 -23 P= .185	0.0113 -23 P= .959
@BLACK	0.0302 -23 P= .891	0.1989 -23 P= .363	-0.6254 -23 P= .001	0.1678 -23 P= .444	0.7538 -23 P= .000	1.0000 -23 P= .	0.4601 -23 P= .027	-0.3785 -23 P= .075	0.4646 -20 P= .039	0.5025 -23 P= .015	0.2854 -23 P= .187
@ONE_HH	-0.0346 -23 P= .875	0.0992 -23 P= .652	-0.6104 -23 P= .002	0.5352 -23 P= .009	0.3395 -23 P= .113	0.4601 -23 P= .027	1.0000 -23 P= .	-0.4002 -23 P= .058	0.3330 -20 P= .151	0.3536 -23 P= .098	0.5842 -23 P= .003
MED_HINC	0.0644 -23 P= .770	-0.0386 -23 P= .861	0.6120 -23 P= .002	-0.3344 -23 P= .119	-0.1635 -23 P= .456	-0.3785 -23 P= .075	-0.4002 -23 P= .058	1.0000 -23 P= .	-0.7546 -20 P= .000	-0.9303 -23 P= .000	-0.5077 -23 P= .013
@HH_ASST	-0.1891 -20 P= .425	-0.0460 -20 P= .847	-0.5974 -20 P= .005	0.6408 -20 P= .002	0.1993 -20 P= .400	0.4646 -20 P= .039	0.3330 -20 P= .151	-0.7546 -20 P= .000	1.0000 -20 P= .	0.7745 -20 P= .000	0.7385 -20 P= .000
@POVERTY	0.0102 -23 P= .963	0.1165 -23 P= .597	-0.6231 -23 P= .001	0.2298 -23 P= .292	0.2867 -23 P= .185	0.5025 -23 P= .015	0.3536 -23 P= .098	-0.9303 -23 P= .000	0.7745 -20 P= .000	1.0000 -23 P= .	0.4440 -23 P= .034
@HU_1939	-0.3886 -23 P= .067	-0.2511 -23 P= .248	-0.6032 -23 P= .002	0.6927 -23 P= .000	0.0113 -23 P= .959	0.2854 -23 P= .187	0.5842 -23 P= .003	-0.5077 -23 P= .013	0.7385 -20 P= .000	0.4440 -23 P= .034	1.0000 -23 P= .

(Coefficient / (Cases) / 2-tailed Significance)

" . " is printed if a coefficient cannot be computed

Independent Variables (continued)

-- Correlation Coefficients --

	POP_1992	POP_1980	@C80_92	@ELDERLY	B_POP90	@BLACK	@ONE_HH	MED_HINC	@HH_ASST	@POVERTY	@HU_1939
@HU_RENT	-0.1572 / -23 / P= .474	-0.0619 / -23 / P= .779	-0.4757 / -23 / P= .022	0.0890 / -23 / P= .686	0.2260 / -23 / P= .300	0.4665 / -23 / P= .025	0.6048 / -23 / P= .002	-0.4592 / -23 / P= .028	0.3831 / -20 / P= .095	0.5595 / -23 / P= .006	0.4940 / -23 / P= .017
UNEMP_RT	-0.2456 / -23 / P= .259	-0.1881 / -23 / P= .390	-0.3258 / -23 / P= .129	0.2953 / -23 / P= .171	-0.0301 / -23 / P= .892	0.1208 / -23 / P= .583	0.0056 / -23 / P= .980	-0.4138 / -23 / P= .050	0.5462 / -20 / P= .013	0.5387 / -23 / P= .008	0.4916 / -23 / P= .017
@C_LABOR	-0.1211 / -23 / P= .582	-0.3061 / -23 / P= .155	0.9805 / -23 / P= .000	-0.6686 / -23 / P= .000	-0.4648 / -23 / P= .025	-0.5695 / -23 / P= .005	-0.6140 / -23 / P= .002	0.6228 / -23 / P= .002	-0.5825 / -20 / P= .007	-0.6301 / -23 / P= .001	-0.5572 / -23 / P= .006
RAINFALL	0.1721 / -23 / P= .432	0.2519 / -23 / P= .246	-0.2987 / -23 / P= .166	-0.0207 / -23 / P= .925	0.4986 / -23 / P= .015	0.5310 / -23 / P= .009	0.0513 / -23 / P= .816	0.0394 / -23 / P= .858	-0.0142 / -20 / P= .953	-0.0047 / -23 / P= .983	0.0450 / -23 / P= .839
PCT_FHFH	-0.1753 / -23 / P= .424	-0.0089 / -23 / P= .968	-0.7338 / -23 / P= .000	0.3647 / -23 / P= .087	0.4520 / -23 / P= .030	0.8026 / -23 / P= .000	0.5906 / -23 / P= .003	-0.6260 / -23 / P= .001	0.7652 / -20 / P= .000	0.7621 / -23 / P= .000	0.6349 / -23 / P= .001
PCT_EDUC	-0.0267 / -23 / P= .904	-0.1544 / -23 / P= .482	0.6831 / -23 / P= .000	-0.3477 / -23 / P= .104	-0.3303 / -23 / P= .124	-0.4998 / -23 / P= .015	-0.1964 / -23 / P= .369	0.7344 / -23 / P= .000	-0.7696 / -20 / P= .000	-0.8594 / -23 / P= .000	-0.4982 / -23 / P= .016
UNDER5	-0.0693 / -23 / P= .753	-0.1479 / -23 / P= .501	0.3638 / -23 / P= .088	-0.5831 / -23 / P= .003	-0.2378 / -23 / P= .275	-0.1765 / -23 / P= .420	-0.4421 / -23 / P= .035	-0.2226 / -23 / P= .307	0.0412 / -20 / P= .863	0.2356 / -23 / P= .279	-0.1819 / -23 / P= .406
BETW5_17	0.1738 / -23 / P= .428	0.0673 / -23 / P= .760	0.4112 / -23 / P= .051	-0.5182 / -23 / P= .011	-0.1612 / -23 / P= .463	-0.2702 / -23 / P= .212	-0.6328 / -23 / P= .001	-0.1889 / -23 / P= .388	-0.0426 / -20 / P= .859	0.2834 / -23 / P= .190	-0.3920 / -23 / P= .064

(Coefficient / (Cases) / 2-tailed Significance)

" . " is printed if a coefficient cannot be computed

Independent Variables (continued)

-- Correlation Coefficients --

	@HU_RENT	UNEMP_RT	@C_LABOR	RAINFALL	PCT_FHFH	PCT_EDUC	UNDER5	BETW5_17
POP_1992	-0.1572 -23 P= .474	-0.2456 -23 P= .259	-0.1211 -23 P= .582	0.1721 -23 P= .432	-0.1753 -23 P= .424	-0.0267 -23 P= .904	-0.0693 -23 P= .753	0.1738 -23 P= .428
POP_1980	-0.0619 -23 P= .779	-0.1881 -23 P= .390	-0.3061 -23 P= .155	0.2519 -23 P= .246	-0.0089 -23 P= .968	-0.1544 -23 P= .482	-0.1479 -23 P= .501	0.0673 -23 P= .760
@C80_92	-0.4757 -23 P= .022	-0.3258 -23 P= .129	0.9805 -23 P= .000	-0.2987 -23 P= .166	-0.7338 -23 P= .000	0.6831 -23 P= .000	0.3638 -23 P= .088	0.4112 -23 P= .051
@ELDERLY	0.0890 -23 P= .686	0.2953 -23 P= .171	-0.6686 -23 P= .000	-0.0207 -23 P= .925	0.3647 -23 P= .087	-0.3477 -23 P= .104	-0.5831 -23 P= .003	-0.5182 -23 P= .011
B_POP90	0.2260 -23 P= .300	-0.0301 -23 P= .892	-0.4648 -23 P= .025	0.4986 -23 P= .015	0.4520 -23 P= .030	-0.3303 -23 P= .124	-0.2378 -23 P= .275	-0.1612 -23 P= .463
@BLACK	0.4665 -23 P= .025	0.1208 -23 P= .583	-0.5695 -23 P= .005	0.5310 -23 P= .009	0.8026 -23 P= .000	-0.4998 -23 P= .015	-0.1765 -23 P= .420	-0.2702 -23 P= .212
@ONE_HH	0.6048 -23 P= .002	0.0056 -23 P= .980	-0.6140 -23 P= .002	0.0513 -23 P= .816	0.5906 -23 P= .003	-0.1964 -23 P= .369	-0.4421 -23 P= .035	-0.6328 -23 P= .001
MED_HINC	-0.4592 -23 P= .028	-0.4138 -23 P= .050	0.6228 -23 P= .002	0.0394 -23 P= .858	-0.6260 -23 P= .001	0.7344 -23 P= .000	-0.2226 -23 P= .307	-0.1889 -23 P= .388
@HH_ASST	0.3831 -20 P= .095	0.5462 -20 P= .013	-0.5825 -20 P= .007	-0.0142 -20 P= .953	0.7652 -20 P= .000	-0.7696 -20 P= .000	0.0412 -20 P= .863	-0.0426 -20 P= .859
@POVERTY	0.5595 -23 P= .006	0.5387 -23 P= .008	-0.6301 -23 P= .001	-0.0047 -23 P= .983	0.7621 -23 P= .000	-0.8594 -23 P= .000	0.2356 -23 P= .279	0.2834 -23 P= .190
@HU_1939	0.4940 -23 P= .017	0.4916 -23 P= .017	-0.5572 -23 P= .006	0.0450 -23 P= .839	0.6349 -23 P= .001	-0.4982 -23 P= .016	-0.1819 -23 P= .406	-0.3920 -23 P= .064

(Coefficient / (Cases) / 2-tailed Significance)   " . " is printed if a coefficient cannot be computed

Independent Variables (continued)

- - Correlation Coefficients - -

	@HU_RENT	UNEMP_RT	@C_LABOR	RAINFALL	PCT_FHFH	PCT_EDUC	UNDER5	BETW5_17
@HU_RENT	1.0000 -23 P= .	0.4257 -23 P= .043	-0.4333 -23 P= .039	0.2142 -23 P= .326	0.7393 -23 P= .000	-0.4758 -23 P= .022	-0.0200 -23 P= .928	-0.2493 -23 P= .251
UNEMP_RT	0.4257 -23 P= .043	1 -23 P= .	-0.3021 -23 P= .161	-0.1816 -23 P= .407	0.4788 -23 P= .021	-0.7341 -23 P= .000	0.089 -23 P= .687	0.2708 -23 P= .211
@C_LABOR	-0.4333 -23 P= .039	-0.3021 -23 P= .161	1.0000 -23 P= .	-0.2291 -23 P= .293	-0.6805 -23 P= .000	0.6830 -23 P= .000	0.3134 -23 P= .145	0.3528 -23 P= .099
RAINFALL	0.2142 -23 P= .326	-0.1816 -23 P= .407	-0.2291 -23 P= .293	1.0000 -23 P= .	0.2666 -23 P= .219	-0.0545 -23 P= .805	-0.1391 -23 P= .527	-0.2860 -23 P= .186
PCT_FHFH	0.7393 -23 P= .000	0.4788 -23 P= .021	-0.6805 -23 P= .000	0.2666 -23 P= .219	1.0000 -23 P= .	-0.7566 -23 P= .000	-0.0942 -23 P= .669	-0.1650 -23 P= .452
PCT_EDUC	-0.4758 -23 P= .022	-0.7341 -23 P= .000	0.6830 -23 P= .000	-0.0545 -23 P= .805	-0.7566 -23 P= .000	1.0000 -23 P= .	-0.1515 -23 P= .490	-0.2442 -23 P= .262
UNDER5	-0.0200 -23 P= .928	0.0890 -23 P= .687	0.3134 -23 P= .145	-0.1391 -23 P= .527	-0.0942 -23 P= .669	-0.1515 -23 P= .490	1.0000 -23 P= .	0.6844 -23 P= .000
BETW5_17	-0.2493 -23 P= .251	0.2708 -23 P= .211	0.3528 -23 P= .099	-0.2860 -23 P= .186	-0.1650 -23 P= .452	-0.2442 -23 P= .262	0.6844 -23 P= .000	1.0000 -23 P= .

(Coefficient / (Cases) / 2-tailed Significance)

" . " is printed if a coefficient cannot be computed

Appendix C.   Correlation Coefficients for the City Characteristics and the Fire Indicators

The pages below contain correlation coefficients for the fire indicators and the 18 city characteristics that were statistically significant with one or more of the fire indicators at the P≤ .05 level

Dependent and Independent Variables

-- Correlation Coefficients --

	POP_1992	POP_1980	@C80_92	@ELDERLY	B_POP90	@BLACK	@ONE_HH	MED_HINC	@HH_ASST	@POVERTY	@HU_1939
FR_NTOTL	-0.0816 -23 P= .711	0.0243 -23 P= .912	-0.4876 -23 P= .018	0.2945 -23 P= .173	0.2384 -23 P= .273	0.2871 -23 P= .184	0.3119 -23 P= .147	-0.3857 -23 P= .069	0.2718 -20 P= .246	0.2832 -23 P= .190	0.4940 -23 P= .017
FR_NARS	0.0590 -23 P= .789	0.1478 -23 P= .501	-0.5534 -23 P= .006	0.3924 -23 P= .064	0.2365 -23 P= .277	0.2772 -23 P= .200	0.5135 -23 P= .012	-0.5988 -23 P= .003	0.4940 -20 P= .027	0.5630 -23 P= .005	0.5708 -23 P= .004
FR_NCHLD	-0.2815 -23 P= .193	-0.1701 -23 P= .438	-0.4989 -23 P= .015	0.0897 -23 P= .684	0.1156 -23 P= .599	0.4599 -23 P= .027	0.3523 -23 P= .099	-0.5716 -23 P= .004	0.4739 -20 P= .035	0.5423 -23 P= .008	0.4588 -23 P= .028
FR_NSMOK	-0.2402 -23 P= .270	-0.1076 -23 P= .625	-0.5879 -23 P= .003	0.3533 -23 P= .098	0.1568 -23 P= .475	0.4156 -23 P= .049	0.5586 -23 P= .006	-0.5003 -23 P= .015	0.3295 -20 P= .156	0.3797 -23 P= .074	0.6271 -23 P= .001
FR_NHEAT	0.0703 -23 P= .750	0.0740 -23 P= .737	0.0132 -23 P= .952	-0.0248 -23 P= .910	0.1454 -23 P= .508	0.1004 -23 P= .649	-0.1149 -23 P= .602	0.0301 -23 P= .892	-0.2965 -20 P= .204	-0.2004 -23 P= .359	-0.1740 -23 P= .427
FR_NCOOK	-0.0777 -23 P= .725	-0.0279 -23 P= .900	-0.1752 -23 P= .424	0.1153 -23 P= .600	0.0786 -23 P= .721	0.0310 -23 P= .888	0.0496 -23 P= .822	-0.0328 -23 P= .882	0.0214 -20 P= .929	-0.0236 -23 P= .915	0.2539 -23 P= .242
FR_NELEC	0.4573 -23 P= .028	0.5417 -23 P= .008	-0.3526 -23 P= .099	0.0457 -23 P= .836	0.6151 -23 P= .002	0.4015 -23 P= .058	0.2441 -23 P= .262	-0.2790 -23 P= .197	0.0794 -20 P= .739	0.2574 -23 P= .236	0.0260 -23 P= .906
FR_NAPPL	0.3890 -23 P= .067	0.4147 -23 P= .049	-0.0933 -23 P= .672	-0.0475 -23 P= .829	0.3222 -23 P= .134	0.0192 -23 P= .931	-0.0206 -23 P= .926	0.0347 -23 P= .875	-0.0911 -20 P= .702	-0.0744 -23 P= .736	0.0064 -23 P= .977
FR_NOPEN	-0.2117 -23 P= .332	-0.1316 -23 P= .549	-0.3292 -23 P= .125	0.1194 -23 P= .587	0.1202 -23 P= .585	0.3288 -23 P= .133	0.3809 -23 P= .073	-0.4196 -23 P= .046	0.1817 -20 P= .443	0.3346 -23 P= .119	0.5206 -23 P= .011

(Coefficient / (Cases) / 2-tailed Significance)

" . " is printed if a coefficient cannot be computed

Dependent and Independent Variables (continued)

-- Correlation Coefficients --

	@HU_RENT	UNEMP_RT	@C_LABOR	RAINFALL	PCT_FHFH	PCT_EDUC	UNDER5	BETW5_17
FR_NTOTL	0.3784 -23 P= .075	0.1080 -23 P= .624	-0.4633 -23 P= .026	0.4833 -23 P= .019	0.3635 -23 P= .088	-0.2634 -23 P= .225	0.1300 -23 P= .554	-0.2796 -23 P= .196
FR_NARS	0.5421 -23 P= .008	0.4707 -23 P= .023	-0.5290 -23 P= .009	0.1712 -23 P= .435	0.5134 -23 P= .012	-0.4746 -23 P= .022	-0.0859 -23 P= .697	-0.1029 -23 P= .640
FR_NCHLD	0.5594 -23 P= .006	0.1658 -23 P= .450	-0.4853 -23 P= .019	0.2112 -23 P= .333	0.6230 -23 P= .001	-0.5224 -23 P= .011	0.3734 -23 P= .079	-0.0599 -23 P= .786
FR_NSMOK	0.5252 -23 P= .010	0.0473 -23 P= .830	-0.5724 -23 P= .004	0.3012 -23 P= .163	0.5090 -23 P= .013	-0.3212 -23 P= .135	-0.0485 -23 P= .826	-0.4304 -23 P= .040
FR_NHEAT	-0.2357 -23 P= .279	-0.4776 -23 P= .021	0.0208 -23 P= .925	0.5360 -23 P= .008	-0.2150 -23 P= .325	0.3080 -23 P= .153	-0.0244 -23 P= .912	-0.1630 -23 P= .457
FR_NCOOK	0.2030 -23 P= .353	0.0001 -23 P=1.000	-0.1605 -23 P= .464	0.3516 -23 P= .100	0.0871 -23 P= .693	-0.0306 -23 P= .890	0.1753 -23 P= .424	-0.2448 -23 P= .260
FR_NELEC	0.1290 -23 P= .558	-0.2251 -23 P= .302	-0.4114 -23 P= .051	0.5288 -23 P= .009	0.2179 -23 P= .318	-0.1847 -23 P= .399	0.0816 -23 P= .711	-0.0419 -23 P= .850
FR_NAPPL	-0.1147 -23 P= .602	-0.2916 -23 P= .177	-0.1516 -23 P= .490	0.3991 -23 P= .059	-0.1003 -23 P= .649	0.0500 -23 P= .821	0.2209 -23 P= .311	-0.0369 -23 P= .867
FR_NOPEN	0.4834 -23 P= .019	0.0876 -23 P= .691	-0.2870 -23 P= .184	0.3119 -23 P= .147	0.5116 -23 P= .013	-0.2681 -23 P= .216	0.1367 -23 P= .534	-0.1517 -23 P= .490

(Coefficient / (Cases) / 2-tailed Significance)

" . " is printed if a coefficient cannot be computed

Appendix D.   Percentage Distribution of Fire Cause by City/County

The pages below contain the percentage distribution of fire cause for each of the 27 cities in the data set.

Percentage Distribution of Fire Cause by City/County

Adjusted Percents

City/County	Incendiary / Suspicious	Children Playing	Careless Smoking	Heating	Cooking	Electrical Distribution	Appliances	Open Flame	Other Heat	Other Equipment	Natural	Exposure
Baltimore City	19%	7%	6%	7%	39%	7%	6%	8%	0%	0%	0%	0%
Baltimore County	11%	5%	7%	14%	23%	10%	13%	6%	3%	1%	1%	6%
Boston	20%	5%	12%	3%	29%	4%	4%	6%	1%	1%	1%	14%
Buffalo	46%	7%	8%	4%	10%	6%	3%	7%	2%	0%	0%	7%
Cleveland	23%	11%	13%	6%	15%	13%	6%	4%	0%	0%	1%	7%
Columbus	21%	9%	8%	7%	27%	9%	7%	6%	1%	1%	1%	5%
Dallas	31%	6%	7%	10%	20%	12%	6%	5%	1%	1%	1%	0%
Denver	18%	8%	8%	6%	33%	8%	6%	7%	1%	1%	1%	2%
Detroit	40%	13%	12%	1%	2%	0%	2%	11%	3%	0%	1%	16%
El Paso	24%	7%	6%	8%	16%	5%	10%	9%	4%	2%	2%	6%
Fort Worth	19%	9%	6%	12%	23%	13%	9%	6%	1%	1%	1%	1%
Houston	15%	3%	5%	9%	33%	16%	10%	4%	0%	1%	1%	2%
Jacksonville	23%	4%	9%	9%	25%	5%	5%	10%	2%	1%	2%	4%
Kansas City	46%	7%	6%	4%	3%	1%	3%	10%	10%	0%	2%	8%
Los Angeles County	26%	8%	2%	6%	11%	7%	9%	10%	2%	6%	1%	12%
Memphis	10%	7%	5%	12%	30%	17%	8%	4%	1%	1%	2%	4%
Montgomery County	15%	4%	7%	10%	25%	14%	10%	9%	1%	2%	4%	0%
Nashville	20%	5%	6%	15%	23%	13%	7%	6%	1%	1%	3%	1%
New Orleans	12%	7%	10%	12%	10%	11%	8%	5%	1%	1%	1%	22%
Norfolk	10%	9%	11%	14%	36%	7%	5%	4%	1%	0%	0%	4%
Oklahoma City	26%	7%	9%	11%	15%	5%	5%	6%	4%	1%	2%	7%
Orange County	13%	1%	1%	6%	29%	11%	11%	2%	1%	20%	2%	3%
Portland	9%	4%	11%	19%	25%	10%	7%	7%	1%	2%	1%	3%
Rochester	10%	9%	8%	5%	40%	7%	8%	6%	5%	1%	1%	1%
San Antonio	30%	7%	4%	8%	22%	8%	7%	7%	2%	1%	2%	4%
San Diego	24%	3%	6%	7%	32%	10%	5%	7%	0%	2%	1%	3%
San Francisco	9%	2%	21%	4%	38%	4%	2%	8%	2%	2%	1%	7%
Virginia Beach	7%	4%	4%	19%	37%	6%	9%	7%	1%	1%	2%	3%
Washington DC	25%	8%	13%	6%	12%	11%	6%	7%	4%	1%	1%	7%
Worcester	19%	3%	5%	5%	49%	4%	6%	4%	1%	1%	1%	4%